MW01383751

HISTORICAL FIGURES

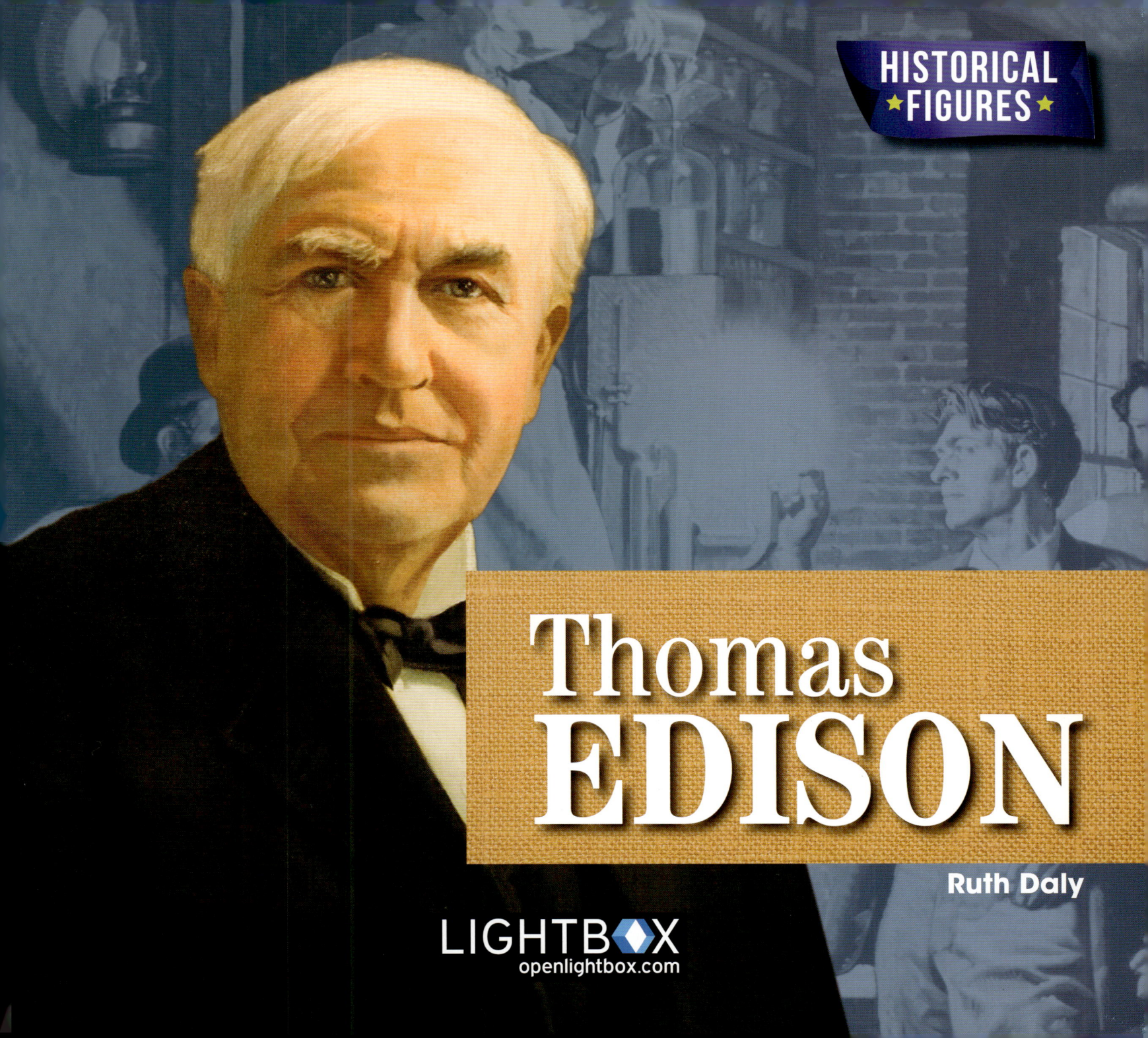

Thomas EDISON

Ruth Daly

LIGHTB✦X
openlightbox.com

LIGHTBOX

Go to
www.openlightbox.com
and enter this book's
unique code.

ACCESS CODE

LBXR5925

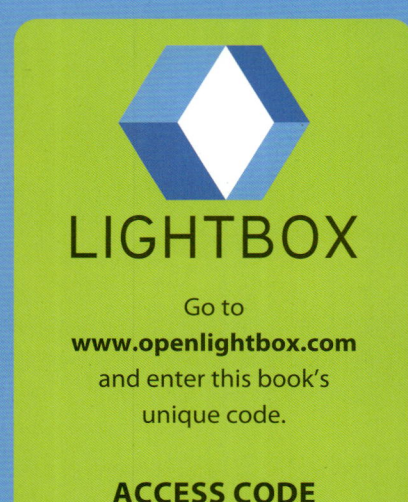

Lightbox is an all-inclusive digital solution for the teaching and learning of curriculum topics in an original, groundbreaking way. Lightbox is based on National Curriculum Standards.

OPTIMIZED FOR
- ✓ TABLETS
- ✓ WHITEBOARDS
- ✓ COMPUTERS
- ✓ AND MUCH MORE!

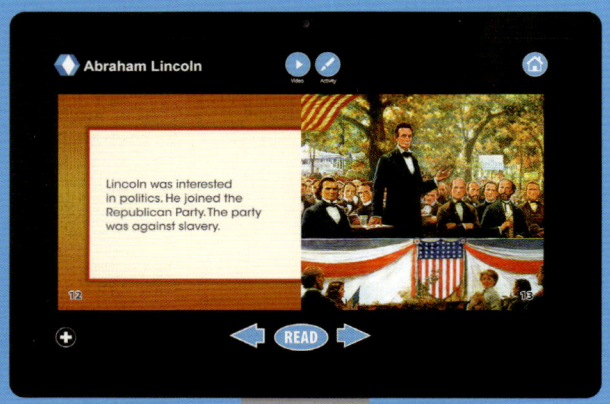

Copyright © 2021 Smartbook Media Inc. All rights reserved.

STANDARD FEATURES OF LIGHTBOX

AUDIO High-quality narration using text-to-speech system

VIDEOS Embedded high-definition video clips

ACTIVITIES Printable PDFs that can be emailed and graded

WEBLINKS Curated links to external, child-safe resources

SLIDESHOWS Pictorial overviews of key concepts

INTERACTIVE MAPS Interactive maps and aerial satellite imagery

QUIZZES Ten multiple choice questions that are automatically graded and emailed for teacher assessment

KEY WORDS Matching key concepts to their definitions

VIDEOS

WEBLINKS

SLIDESHOWS

QUIZZES

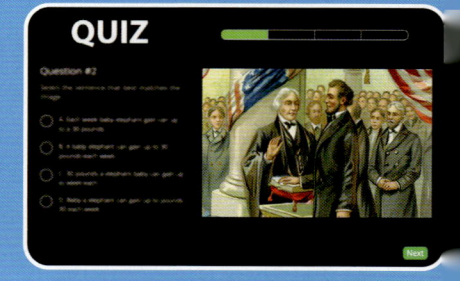

Thomas EDISON

Contents

- 2   Lightbox Access Code
- 4   Who Was Thomas Edison?
- 6   Growing Up
- 8   Overcoming Obstacles
- 10 Values
- 12 Science
- 14 Thoughts from Edison
- 16 Inventor
- 18 Achievements and Successes
- 20 Thomas Edison Timeline
- 22 Cause and Effect
- 24 Key Words

Thomas Edison was an inventor. One of his most important inventions was a new kind of electric light bulb.

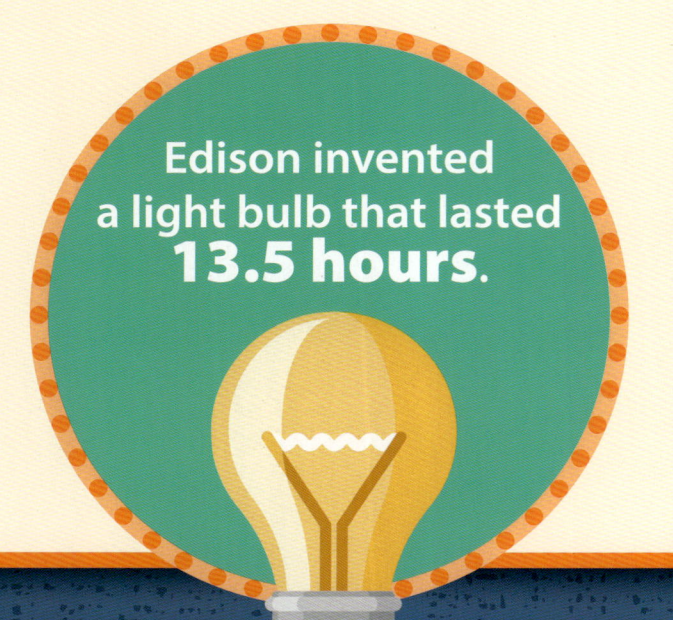

Edison invented a light bulb that lasted **13.5 hours**.

5

Edison was born in Ohio. His family moved to Michigan. Edison grew up there.

Edison's family home in Ohio is now a museum of his life.

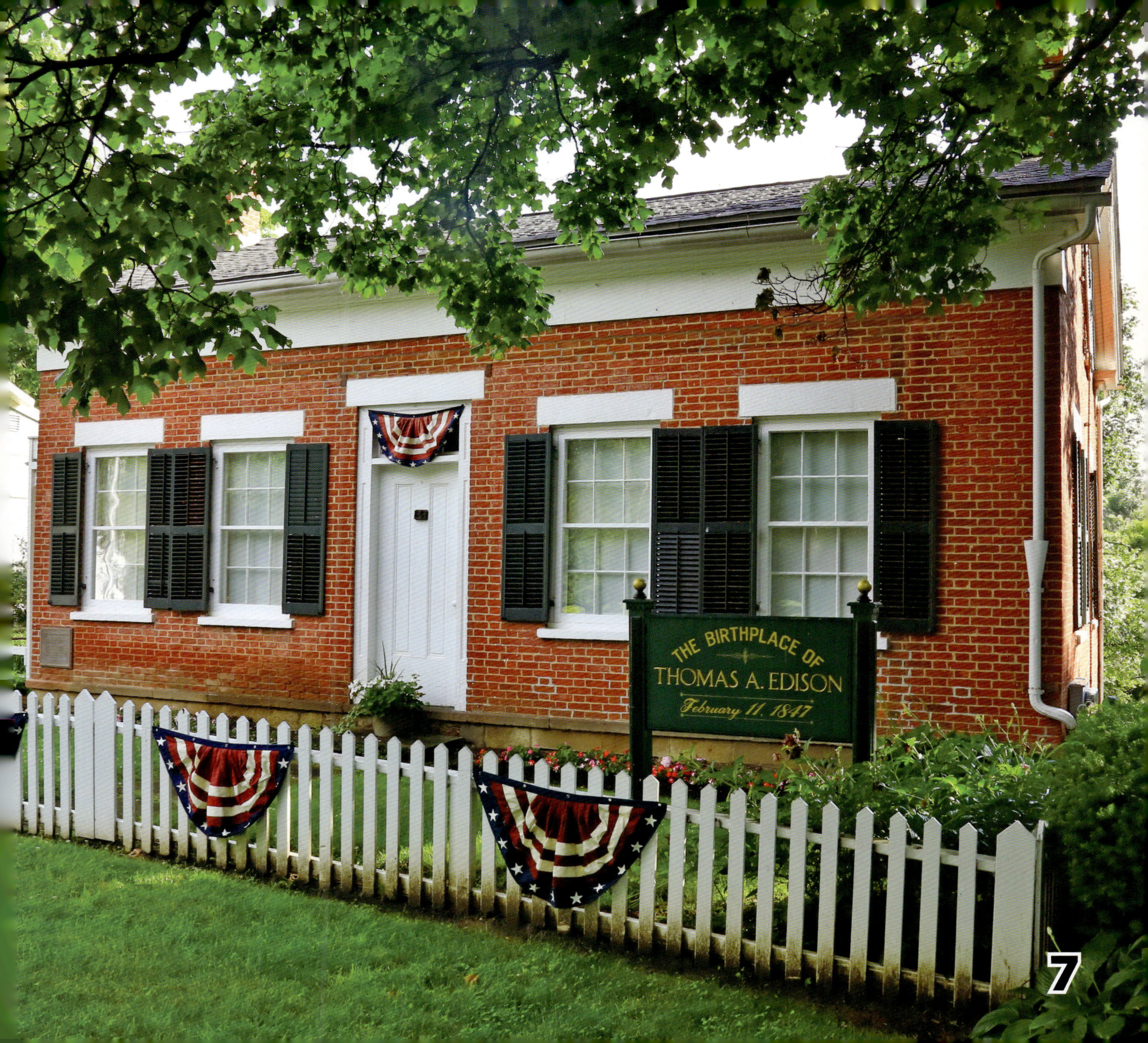

7

Edison could not hear well. He had trouble at school. He learned by reading many books.

9

Edison believed in hard work. His first job was on a train when he was 12. Sometimes, he worked all night.

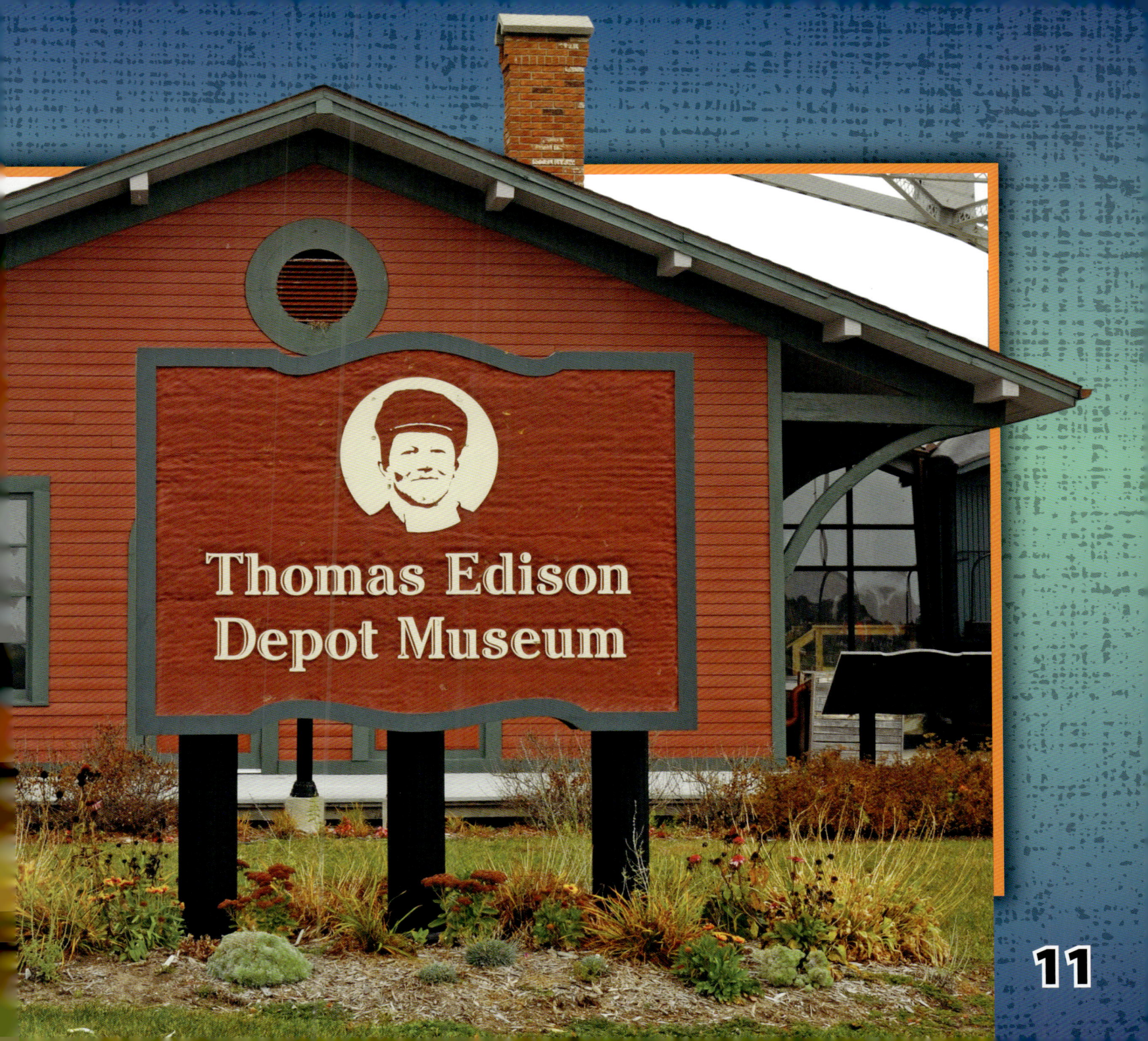

Thomas Edison
Depot Museum

11

Edison was interested in science. He liked finding out how things worked.

He built his own laboratory. This was where he did experiments.

13

Edison enjoyed his job. He thought it was important for people to like their work.

"I never did a day's work in my life. It was all fun."
– Thomas Edison

15

Edison became an inventor. Workers called muckers helped him in his laboratory.

Edison invented more than **1,000** different things.

17

Edison helped bring electricity to the world.

He also invented the phonograph. It recorded sounds.

The **Thomas Edison Memorial Tower** in **Menlo Park, New Jersey,** is 134 feet tall.

19

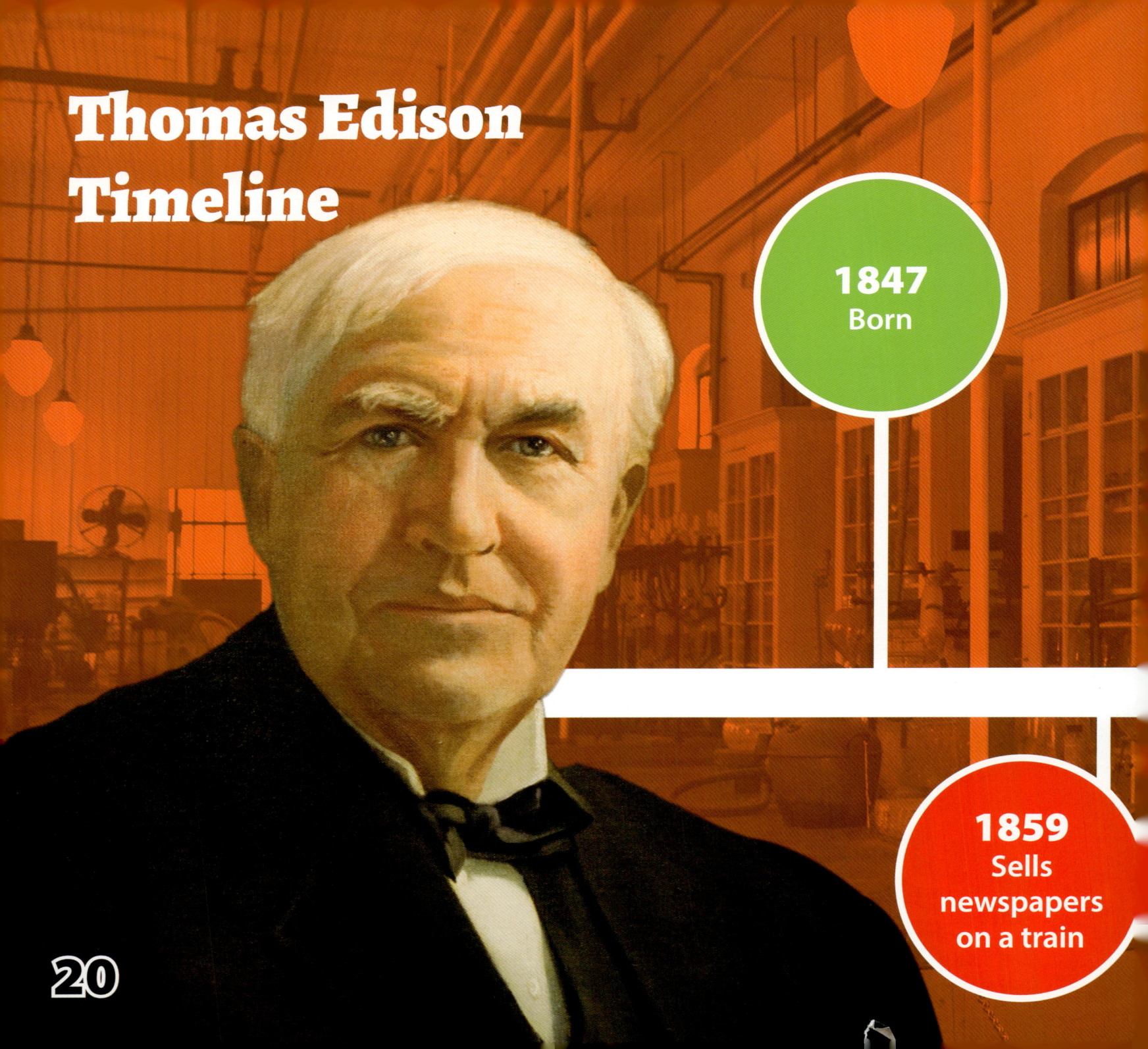

Thomas Edison Timeline

1847
Born

1859
Sells newspapers on a train

20

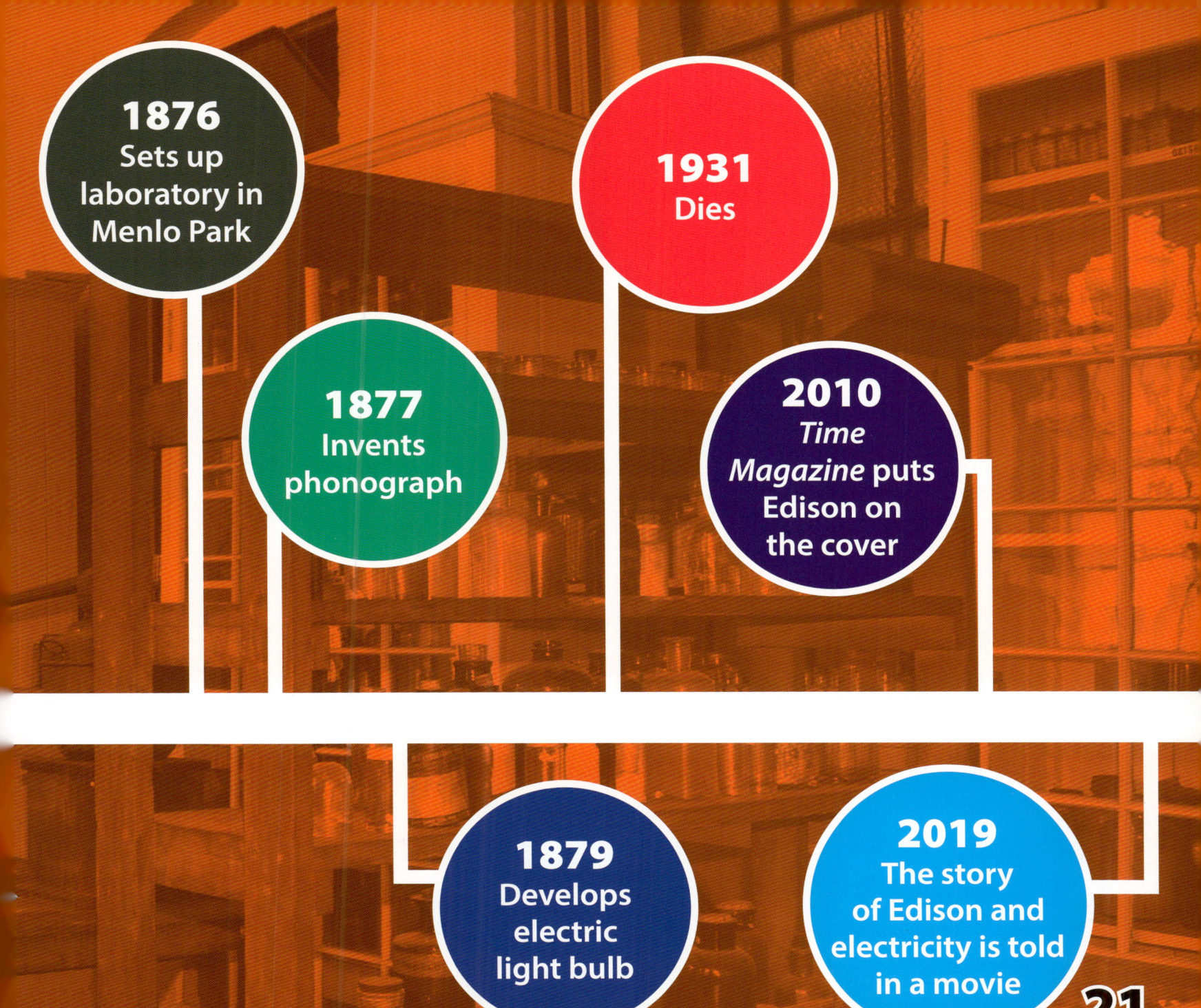

- **1876** Sets up laboratory in Menlo Park
- **1877** Invents phonograph
- **1879** Develops electric light bulb
- **1931** Dies
- **2010** *Time Magazine* puts Edison on the cover
- **2019** The story of Edison and electricity is told in a movie

Cause

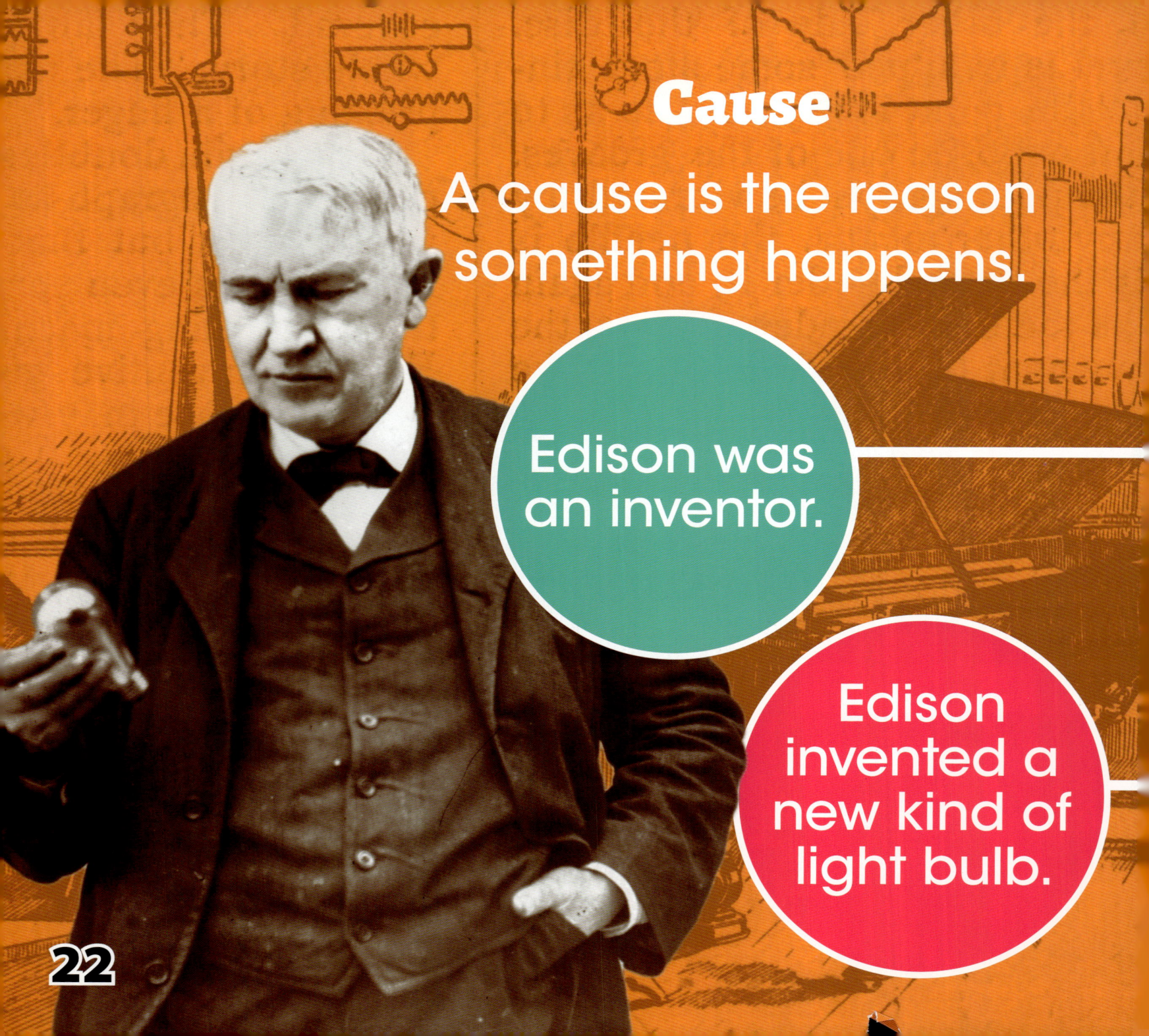

A cause is the reason something happens.

Edison was an inventor.

Edison invented a new kind of light bulb.

Effect

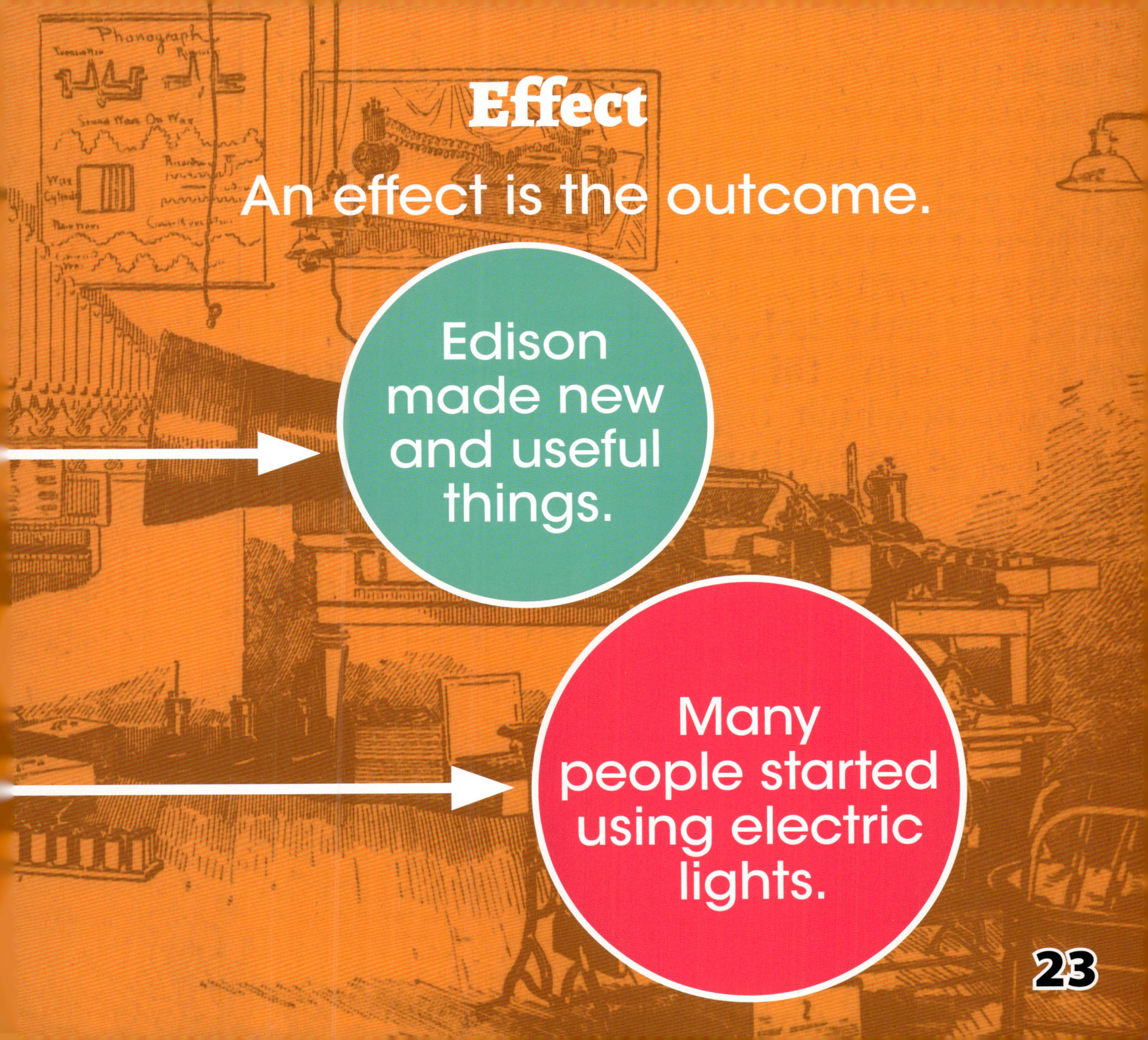

An effect is the outcome.

→ Edison made new and useful things.

→ Many people started using electric lights.

KEY WORDS

Research has shown that as much as 65 percent of all written material published in English is made up of 300 words. These 300 words cannot be taught using pictures or learned by sounding them out. They must be recognized by sight. This book contains 74 common sight words to help young readers improve their reading fluency and comprehension. This book also teaches young readers several important content words, such as proper nouns. These words are paired with pictures to aid in learning and improve understanding.

Page	Sight Words First Appearance
4	a, an, his, important, kind, light, most, new, of, one, that, was
6	family, home, in, is, life, now, there, to, up
8	at, books, by, could, had, he, hear, many, not, school, well
10	all, first, hard, night, on, sometimes, when, work
12	did, how, out, own, things, this, where
14	day, for, I, it, like, my, never, people, their, thought
16	different, him, more, than
18	also, sounds, the, world
19	feet
21	and, puts, sets, story, time
22	something
23	made, started

Page	Content Words First Appearance
4	bulb, electric, hours, inventions, inventor, Thomas Edison
6	Michigan, museum, Ohio
8	trouble
10	job, train
12	experiments, laboratory, science
14	fun
16	muckers, workers
18	electricity, phonograph
19	Menlo Park, New Jersey, Thomas Edison Memorial Tower
20	newspapers, timeline
21	cover, movie, *Time Magazine*
22	cause, reason
23	effect, outcome

Published by Smartbook Media Inc.
350 5th Avenue, 59th Floor New York, NY 10118
Website: www.openlightbox.com

Copyright ©2021 Smartbook Media Inc.
All rights reserved. No part of this publication may be reproduced, stored in a retrieval system, or transmitted in any form or by any means, electronic, mechanical, photocopying, recording, or otherwise, without the prior written permission of the publisher.

Library of Congress Cataloging-in-Publication Data

Names: Daly, Ruth, 1962- author.
Title: Thomas Edison / Ruth Daly.
Description: New York : Lightbox, 2020. | Series: Historical figures | Audience: Ages 4-8 | Audience: Grades K-1
Identifiers: LCCN 2020014161 (print) | LCCN 2020014162 (ebook) | ISBN 9781510553781 (library binding) | ISBN 9781510553798 | ISBN 9781510553804
Subjects: LCSH: Edison, Thomas A. (Thomas Alva), 1847-1931--Juvenile literature. | Inventors--United States--Biography--Juvenile literature. | Electrical engineers--United States--Biography--Juvenile literature.

Classification: LCC TK140.E3 D35 2020 (print) | LCC TK140.E3 (ebook) | DDC 621.32/6092 [B]--dc23
LC record available at https://lccn.loc.gov/2020014161
LC ebook record available at https://lccn.loc.gov/2020014162

Printed in Guangzhou, China
1 2 3 4 5 6 7 8 9 0 24 23 22 21 20

042020
110819

Project Coordinator: Priyanka Das
Designer: Ana María Vidal

Every reasonable effort has been made to trace ownership and to obtain permission to reprint copyright material. The publisher would be pleased to have any errors or omissions brought to its attention so that they may be corrected in subsequent printings.

The publisher acknowledges Alamy and Getty Images as the primary image suppliers for this title.